Welcome to My Farm

Agri-tourism at its Best

17 Ways to Make Money From Your Farm

By
Darla Noble

Country Life Books

JD-Biz Publishing

Table of Contents

Introduction

The word 'agri-tourism' is a word coined by the agricultural (farming) industry to define what takes place when someone experiences the happenings of a particular farm by way of visiting and possibly participating in the workings of that farm. It's a form of educational entertainment.

There's something else you need to know about agri-tourism, though…well, actually *t*wo something's: 1) agri-tourism can work for almost every farm in some way, shape or form. 2) agri-tourism is virtually limitless but requires you to think outside the box and beyond the normal mindset and perceptions of what farming is.

Oh, yes, and one more thing…agri-tourism can be the way in which you financially enhance your farm's profitability.

Chapter 1 How to Use this Book

This book is meant to be multi-purposeful. It is meant to:

1-Spark your creative juices in order for you to make your farm more profitable. Within the next several pages you will find a number of ways to utilize your resources to bring knowledge, entertainment and goods and services to the public.

2-Provide you with answers to most of your questions regarding the technical end of agri-tourism. We will cover such topics as insurance, liability, bookkeeping, licensing and permits and more—you know, the 'fun' stuff.

Income	Expense-type	Date	Amount

3-Offer advice in regards to where to go for help in starting and promoting your agri-tourism venture. Getting off to a positive start saves stress, time, money and energy. And once you've gotten started, your success largely depends upon how good you are at letting people know what you have to offer. Marketing can be fun and it must be done.

NOTE: This book is <u>not</u> meant to serve as legal counsel or as a substitute for professional tax preparation and filing laws and regulations, but rather as an informational guide.

Chapter 2: The List

This chapter is about as straight-forward as it comes. It consists of a list of agricultural (and some not-so-agricultural) ventures and how YOU can put an agri-tourism spin on them.

FYI: Some of what you read can and does fall under the concept of value-added agriculture as well as agri-tourism. But that's a subject for another book. Oh, yah, we've already written that one (***INSERT NAME OF VALUE-ADDED BOOK HERE)

Bee keeping—
While you will not be able to maintain and support an agri-tourism business with one or two hives, you don't need to have more than a half-dozen to successfully operate a small to mid-sized agri-tourism business.

Bees working in the hive
As a beekeeper, you can invite people onto your location to observe the workings of the bees, to learn via 'artificial' hives how honey is made (the overall process) and they can work with beeswax or create honey straws or other edible or useful products. And as is the case with EVERY agri-tourism business, there needs to be an educational element for children; coloring pages, puzzles, worksheets, etc.

Beef cattle producer—

Farmers whose primary focus is on the beef industry will find that their best choices for agri-tourism will be found in the areas of farm tours and educational stations (especially for children).

These could include touring the barn, learning what tools are required for working with cattle (working chutes, ear taggers, scales, etc.) and also models of beef carcasses to learn how and why meat is graded.

Berry /nut farming—

The possibilities of agri-tourism when it comes to nuts and berries is nearly limitless. U-pick operations are the obvious choices, but don't let your vision stop there. A kitchen devoted to teaching people to make jams and jellies, nut butters and nut candies is also well-received by people wishing to learn something new.

Fresh-picked Raspberries and Blackberries

Chicken/turkey producer—

It never ceases to amaze me that so many children (and even adults) don't know where their food comes from. They just assume it magically appears on the shelves of your local grocery store or super store. So when you raise any form of livestock, you have both the opportunity and responsibility to let people know that the saying, "No farms...no food" is true.

Providing the experience of seeing an incubator full of eggs, clipping wings, sexing chicks, gathering eggs and seasonal experiences such as THE Easter event of the region...a HUGE egg hunt (or private hunts for groups) complete with egg decorating.

A quality digital camera and a few chicks against a simple backdrop will also make you popular with parents who want pictures of their children in this type of setting. As for the educational element, you can offer egg recipes, chicken recipes and worksheets and coloring sheets for kids.
If you prefer to talk turkey, having seminars on how to cook turkey will always be popular.

Last but not least, never underestimate the power of offering group tours through your poultry barns; tours that show and tell the process from hatched egg to being market-ready.

Christmas tree farming—

Christmas time is especially joyous in the agri-tourism industry.

The Christmas tree industry is one that struggles in some areas due to the increasing popularity of the not-so-messy artificial tree. So while you may see a decrease in your cut-your-own tree sales, there are other ways to draw people in.

Cutting fresh greenery from the trees for wreath-making classes makes excellent use of your product, as does offering the experience for children to make and decorate simple dough ornaments or ornaments made from pine or cedar 'coins' (quarter-sized piece of wood cut ¼ to ½ inch thick).

An ornately-decorated tree on your farm complete with an old wagon, rocking chair, or other fun pieces will make the perfect setting for family Christmas card photos. If you have a good camera you can do it, or work on percentages with a photographer you know and trust.

For those who still desire to cut their own tree, be sure to make the experience more special and memorable by offering hot chocolate, playing Christmas music and handing out small fliers that give the history of the Christmas tree.

Crop farming (grains)—
The reality of the situation with crop farming is that it is usually done on such a large scale that there's little time left for anything else. But that doesn't mean someone cannot build an agri-tourism business around a crop or crops rather than growing commercially. For

example: Frank has five acres he grows sorghum on each year. He grows it specifically for guests to spend time each fall during designated times to press the sorghum into molasses the old-fashioned way (horse-drawn mill and cooking it down in large iron kettles over and open fire). Not only do people pay to participate, but others pay to watch the event and enjoy old-fashioned music, food and old-time games.

Sorghum isn't the only crop you can do this with. After combining wheat, farmers who want to include an agri-tourism element in their farm often have horse-drawn straw baling for visitors to experience or scarecrow making events. Those who raise corn sometimes have the very popular corn maze adventure.

Dairy farm—

Dairy farmers can offer the experience of making butter, ice cream, watching the cows being milked, learning what it takes to get the milk from the cow to the store shelves and possibly even try their hand at milking.

Other options include classes in cheese making, ice cream making, and learning to make cottage cheese and whipped cream.

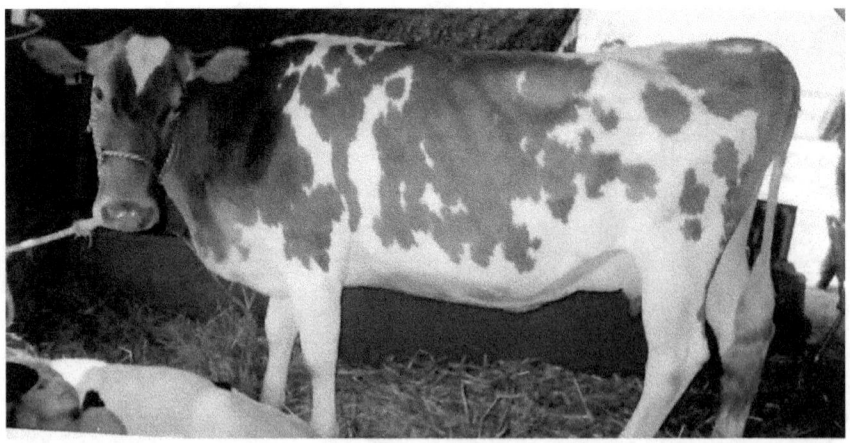

Kandy the Guernsey gives rich, sweet milk.

Some of the things you can make from milk.

Exotic animals—

Those of you with a desire and flare for raising exotic animals often consider a petting zoo to be your best, if not only, option for agri-tourism.

Not true! In fact, while petting zoos are lots of fun, the costs of having one (insurance/liability) are often so high it becomes impossible to maintain it. Instead, opt for educational stations—look but don't touch. There's much to be learned about the eating habits, life cycles and the roles these animals play in nature.

The hunting aspect of agri-tourism can lean toward the traditional (deer, elk, trophy horns or antlers) OR if can be more old-school by offering fox hunting, birds (ducks, quail, etc.)—including the dogs and trapping (catch and release with safe traps that don't harm the animal).

Fish farming—

This one is pretty cut and dried, but instead of just being a place to fish, be the host for sponsored fishing tournaments for kids and dads/kids. Rent fishing poles and tackle, teach people how to create a positive environment for fish in their ponds and become a source for purchasing fingerlings for stocking small ponds.

Fish farmers often find great satisfaction and success in reaching out to civic groups wishing to teach trouble young people to enjoy themselves in a relaxing and constructive manner. You can also contact your local juvenile authorities, after-school programs and youth homes to let them know what you have available.

A happy fisher(wo)man with quite a catch!

Flower and/or herb gardening—
This is another agricultural venue in which the possibilities for agri-tourism are wide and varied. And like bee keeping and a few others, you don't have to have acres and acres of flowers or herbs to make a go of it.

Do you have exquisite iris beds or roses? Are your flower beds a mecca for every butterfly around? Set hours for touring your flower beds and listening to a bit of flower-lore along the way appeals to a great number of people.

Selling plants and seeds on-site is popular, as well. Other options include teaching classes in flower arranging, dried flower crafts and container gardening. You can do this in an open-air lean-to or a barn you've 'converted' into a potting shed. Any or all of these can bring money and marketing opportunities to your farm.

Seed kits sold in classes held at a local greenhouse.

Greenhouse—

Having a greenhouse allows for visits not only for the purchase of buying plants, but also gives you the possibility of teaching gardening classes and educating young people on how things grow. There are even a few greenhouses that 'rent' space to people for the winter months; allowing them to winter potted plants they cannot take inside for the winter.

NOTE: This requires a thorough cleaning and de-bugging of plants for the sake of everyone.

Touring a popular greenhouse

Horses—

Riding lessons and or mini trail rides are popular with a large percentage of city-dwellers. Just be sure you have the ability to put people at ease about riding if they aren't sure, as we all know what effect nervous people can have on a horse.

If you have three or four gentle and well-broke ponies, horses or donkeys, birthday parties can be a source of significant income for you—even more so if you have an outbuilding suitable for parents to serve cake and ice cream (not provided by you) and for opening gifts.

Celebrating a birthday on the back of a miniature donkey.

Learning to ride safely and enjoyably

Naturally resourced (ponds, caves, streams, etc.)—
If you are fortunate enough to have natural landscapes that lend
themselves to potential income opportunities, you don't need to be

linked to the agricultural community in order to take advantage of what is considered agri-tourism.

Do you have a SAFE cave on your property? Are you willing to do what it takes (if anything) to become qualified in leading people through it?

Do you have a picture-perfect setting and/or a structurally-sound barn you aren't using that could be used for weddings, family reunions, or photo sessions?
Do you have a wooded area you could offer up as a location for children's nature hikes and hunts? Can you cultivate mushrooms in this area for hunting?

Orchards/vineyards—

U-picks are also a natural form of agri-tourism with these types of agriculture, but there are other options to consider in place of or in addition to u-pick farms. Some of these include: grape stomping or wine making or taste-testing venues, jelly making classes, fruit drying classes or something of that nature.

The fare from a u-pick apple orchard

Retired farmers—

Wait! Why would a retired farmer be interested in agri-tourism opportunities? To retain their connection in the agricultural community, that's why. There comes a time when it becomes impractical, if not unsafe for the famer to be working cattle, operating heavy machinery or hauling hay. But that doesn't mean they have to hang up their hat and sit idly by while everyone else keeps working.

The retired farmer can take advantage of agri-tourism opportunities by:

1—Hosting hayrides in the fall.

Let's go on a hayride!

2—Renting out the barn for birthday parties, family reunions, etc.
3—Hosting classes on farm management.

Older farmers have much to offer in the way of wisdom and experience.

4—Hosting 4-H and FFA classes in farm safety, proper operation of machinery and crop or livestock production.

In other words…share your wealth of knowledge with the next generation of the agricultural community.

Sheep Producers—

I don't know anyone who can resist a sweet little lamb, so farm tours are a great way to add profitability to your farm. And because sheep/lambs are easy to handle, letting young visitors (and old ones, too) pet a sweet little lamb is never a baaaaaad idea.

For those who produce wool sheep, your agri-tourism experience could involve cleaning wool, dying the wool different colors and possibly even carding and spinning. In fact, you might even specialize in experiences for spinners (beginners and the more advanced).

Touring the lambing barn

Tunis and Suffolk ready to be sheared

For meat sheep producers, lamb cooking classes are also an option.

Even though sheep production has gained a good amount of much-deserved respect over the last fifteen to twenty years, proper sheep management is somewhat of a mystery to many, so those of you who have a firm grip on what it takes to raise and market sheep successfully and profitably can share your information with others—especially new producers and young/future sheep producers (4-H and FFA).

Vegetable gardening—

Vegetable gardeners can take advantage of agri-tourism by focusing on one or two products (example: watermelon, pumpkins/gourds, or herbs and greens) for a u-pick operation OR they can have a roadside stand offering all the bounty summer has to offer our appetites. But that's not the only way to take advantage of agri-tourism dollars…

Some large-scale vegetable farmers charge nominal fees for visitors to experience harvesting crops such as popcorn, sweet corn, beets, potatoes and others that are harvested at one time. Not only do they get the experience of the work itself, but it is followed by classes in cooking and preserving.

With the push to eat healthier and 'greener', home-grown produce is once again becoming something more people wish to do. But wishing and actually being able to do so are often two very different things.

A community garden or co-op garden is great way to offer an agri-tourism experience and make your garden(s) more profitable. You can do this by allowing participants to rent garden space or work in the gardens in exchange for produce.

Fresh produce…it's all good.

Well, are you ready to get started? Keep those ideas coming and don't let your enthusiasm waiver, but keep reading because there are a few things you need to know so you don't get the cart before the horse. Or count your chicken before they hatch. Or…you get the point.

Chapter 3: Covering all the Bases

Getting involved in agri-tourism can and should be fun, enjoyable and profitable. You shouldn't have to be worried or bothered by financial hassles, legalities of being a legit business and the fear of someone suing you. And you don't have to be as long as you cover all the bases right up front.

Safety.

As you prepare your farm for visitors, you need to take whatever safety precautions are necessary to avoid accidents and injuries. Posting signs can often help. For example, if there is a step up or down people might miss, post a sign that says WATCH YOUR STEP. Bees sting. So, whether or not its beekeeping or gardening, these little critters are going to be around. Therefore, it would be wise to post signs that say BEES AT WORK. Do you have barbed wire fencing? Are you pulling a hay wagon with horses? Post caution signs stating that the fence is sharp and that the horse might bite.

Other signs you need to post regardless of what you do include: NOT RESPONSIBLE FOR ACCIDENTS, DO NOT CLIMB ON FENCES AND GATES, DO NOT FEED THE ANIMALS, and RULES FOR PARTICIPATION.

Other matters of safety you must take into consideration include:

*Making sure no keys are left in machinery and that all livestock is secured and out of harm's way or cannot harm visitors.

Keep equipment safely secured

*Keep all tools, hoses, etc. up off the ground.

*Make sure you have ample parking for visitors that is safe and that doesn't cause a traffic hazard.

*If you have a u-pick operation, you will need to make sure moles, gophers or other critters don't dig holes in the ground that will create walking hazards for your patrons.

Legal 'stuff'
Once you decide to include agri-tourism in your farming operation, you need to take the proper steps to protect your personal assets. For most, this means setting your farm up as an LLC (limited liability corporation). An LLC keeps your personal assets (home, savings/checking account, etc.) safe from being included in a legal settlement should one ever occur. Others, however, are content to set

their farm up as a sole proprietorship (husbands/wives count as one person (sole)). If your agri-tourism experience is extremely low-risk, this may be sufficient.

The choice is yours.

Also at the top of the list on the legal side of things is obtaining the proper licenses and permits necessary to do business.
If there is no cooking involved, the only license or permit you will probably need is a county business license. This can be obtained from your county collector. A license usually doesn't cost more than $25 to $50 per year.

If you are teaching any sort of cooking classes, you will most certainly need to hold these classes in an approved kitchen; approved by your county's health department. This kitchen will need to be separate from your own kitchen—even separate from your own home (or located in the basement or an attached room with a separate entrance.

Along with getting your licensing in place, obtaining adequate insurance is essential. Most farm policies include a certain amount of liability, but if you are going to have a regular flow of people coming and going, you need to make sure you have a liability umbrella (policy) in place to protect your assets. As you know, liability insurance is basically betting…that something won't happen. This makes it relatively inexpensive—especially considering the alternative.

Ownership, licensing, insurance in place…check! Next on the list are two items your insurance will either require or recommend. The two items are a) signage b) liability waivers/permission slips.

We've already mentioned signage when we talked about safety, but it's worth mentioning again. NOT RESPONSIBLE FOR ACCIDENTS and KEEP OUT (of off-limit areas) are appropriate for any agri-tourism location.

Liability waivers or permission slips are a must when you are dealing with children—especially if your venue involves direct contact with animals, equipment or high activity (petting zoos, riding stables, hay rides, nature hikes, etc.). These waivers needn't be complex or wordy. A simple statement of excusing you from all responsibility in the case of accident or injury combined with a statement saying you will provide a safe environment and experience to the best of your ability. The waiver should also list all rules and fees applicable to your farm experience.

While the agri-tourism experience should be fun for you as well as your visitors, record-keeping is one of those things that can make or break it for YOU.

Not only must you keep track of all income and expenses, you will need to obtain a tax number from your state's department of revenue. This will allow you to purchase supplies used for your agri-tourism business tax-free. This is done because you will need to charge tax on the goods you sell to visitors to your farm. Income taxes will be paid on a quarterly basis. Be sure you take this into consideration when charging visitors and make sure you take the necessary steps to file correctly.

If, however, yours is a service-oriented business, this tax number may not be necessary. Hayrides, petting zoos, family reunions or weddings, horseback riding lessons…these things don't require a tax number. The supplies needed for these types of business ventures (feed, fertilizer, etc.) aren't taxed at the feed store if you are on record for filing a Schedule F with the IRS.

As for income and expenses, you can use the expense section of the Schedule F (IRS) to itemize your expenses in your bookkeeping system. Think about it…if you have expenses (and income, for that matter) broken down as it needs to be for doing your taxes, getting them done will be much easier. A central location for all receipts (income and expense) will make it easy for you to enter the amounts into your system on a weekly or monthly basis (depending upon your business).

Getting the essential legalities in place before you begin taking visitors onto your farm is a must. Keeping track of your income and expenses is equally important. But don't let it bog you down. Do what needs to be done, get a system in place that works well for you and stick with it.

Chapter 4: Welcome to the Farm

Remember that famous line from *Field of Dreams*? You know the one, "If you build it, they will come". Wouldn't it be nice if it was really that easy?

I've always said that someone could have the best thing since sliced bread, but if no one knows about it, they've got nothing. In other words, marketing and advertising are essential elements of a successful agri-tourism business.

Marketing and advertising
your agri-tourism experience doesn't have to be expensive but it does need to reach the masses as well as being strategically targeted to reach your most likely clientele. You can reach both the masses and specific groups by:

*Putting magnetic signage on your vehicles. Make sure to include a name, phone number and website.

*Joining your state's agri-tourism association. These associations provide mass exposure and marketing for very little cost. If your state doesn't have such a program, contact the state department of agriculture to see what other options you might have.

*Printing (or have them printed) fliers or brochures to place in your local chamber of commerce or visitor's center.

*Purchasing business card-sized ads to sponsor FFA , 4-H and county fairs and/or purchase trophies for youth livestock shows.

Farm-sponsored livestock trophy

*Having t-shirts and hats made with your name/logo on them and wearing them proudly.

*Participating in local and regional farm expos.

Making sure your farm is represented at local, regional, state and even national farm shows

*Send letters/emails/fliers (one or a combination of) to preschools, schools, play groups, churches, civic groups, children's clubs/organizations (scouts, etc.) and college fraternities and sororities.

*Place posters or business cards on bulletin boards.
*DEVELOP A WEBSITE! The age of the internet makes this one a MUST! Oh, and keep it current.
One of the most commonly asked questions to come up when getting started in the agri-tourism business is, **"How much should I charge?"** Ultimately it is your choice, but when deciding how much to charge, you need to take into consideration the following:

www.yourfarmname.com

Dollars and cents

*Is the cost reasonable for most everyone? For example: Charging $3-5 per person for a farm tour which lasts an hour may not sound like much, but in reality, that's about the max schools will ask parents to pay for something like that.

*Will the amount you charge cover your time and expenses? Let's go back to the farm tour 'gig'. The livestock is going to be there whether you are giving tours or not, as is the fact that you will be feeding and caring for them. The cost of doing so is part of farming. So all you are really out is your time and hopefully the cost of a simple coloring book or fun pages about (cows, sheep, or whatever). So by charging $3 per person ends up grossing you (on average) $60 for an hour or so of your time. This is based upon the fact that the average group size is 20 children per class.

*What are visitors getting for their time spent on the farm? U-pick operations usually charge by the pound of fruit/veggies/berries sold. That's easy enough to see. Other agri-tourism experiences which are more service-oriented, however, will require you to put a bit more thought into it. Example: Hayrides. Are you just pulling a hay or straw-laden wagon with a tractor or horse? Or are you providing a place for roasting hotdogs and marshmallows? Not only do you need to make sure your cost is easy on people's budgets, but you also need to take into consideration the cost of insurance and extra trash service that might be necessary, the cost of wood used to build fires for cooking out, etc.

Over all, the cost of most farm tours tends to run between $3-5, classes for the skills discussed in chapter two usually run about $10 per session plus supplies, u-pick operations are based on price per pounds purchased, greenhouses and on-site farm stands price products by the pound and individually and service experiences run on the average of $5-$15 per person depending upon what you offer and where you live.

More than just a trip to the farm—

that's what people are looking for. Like just about everything else in society, your visitors will want to know what's in it for them. They want to have something to take with them to remind them of their experience. So whether it be activity pages, seedlings, pencils, refrigerator magnets with your farm logo, jelly, flowers, food, or even a fist-full of freshly dyed wool, people want something to remember you by. So make sure you give it to them.

FYI: The little something visitors take with them is often your best advertisement.

Service with a smile.

Trust me when I say you are going to get all sorts of people. There will be those who really want to learn, some who think what you do is either disgusting or cruel and even those who think they know more about what you do than YOU do…because they read it in a book or saw it on television.

But when all is said and done, these people are your guests…they are paying guests, and deserve your kindness, attention and respect.

Okay, I can't let this go without sharing with you a TRUE story about one of the farm tours I gave several years ago. A group of first graders came to tour the lambing barns and greenhouse. They were a great group, but one of the mothers who came as a parent helper was somewhat of a nuisance.
She asked questions such as "How can you stand to murder those sweet little lambs?" and "Don't you ever get tired of smelling poop?"

As they were getting ready to leave, our outdoor wood furnace kicked on—causing smoke to pour out of the 'chimney'. As soon as it did she turned to me and asked, "What are you cooking?" I looked at her questioningly and she repeated herself, this time pointing at the furnace. "I'm not cooking anything," I said. "That's the furnace that heats our house, our water and the greenhouse."

She then looked at me like I had two heads and said, "Well, that's dumb. Why not just have it inside the house like everyone else?"

I just smiled and said, "I think they're waiting for you on the bus." Then I turned to wave goodbye to the children.

Children saying goodbye to one of the lambs at G5Farm

Closing Remarks

Agri-tourism can and should be a rewarding experience for everyone. And it will be as long as you take the necessary steps to protect yourself, your guests, your property and your livestock.
Agri-tourism is also the key to ensuring that farming retains its rightful place at the top of the list of what's important to the health, well-being and continuation of society. So with that being said, get out there and show the world how much you enjoy being who you are...an American Farmer.

Author Bio

Darla Noble is a native of mid-Missouri where she lives with her husband of thirty-three years, John. Darla's love of writing began in the fourth grade; after meeting up and coming children's author, Judy Blume,
who, by the way, autographed Darla's copy of "Are you there, God...it's me, Margaret".

Darla's love for writing and family makes her work sought after in the Christian market, parenting and family resources and ghostwriting for educators and inspirational speakers.

Check out some of the other JD-Biz Publishing books

Gardening Series on Amazon

Health Learning Series

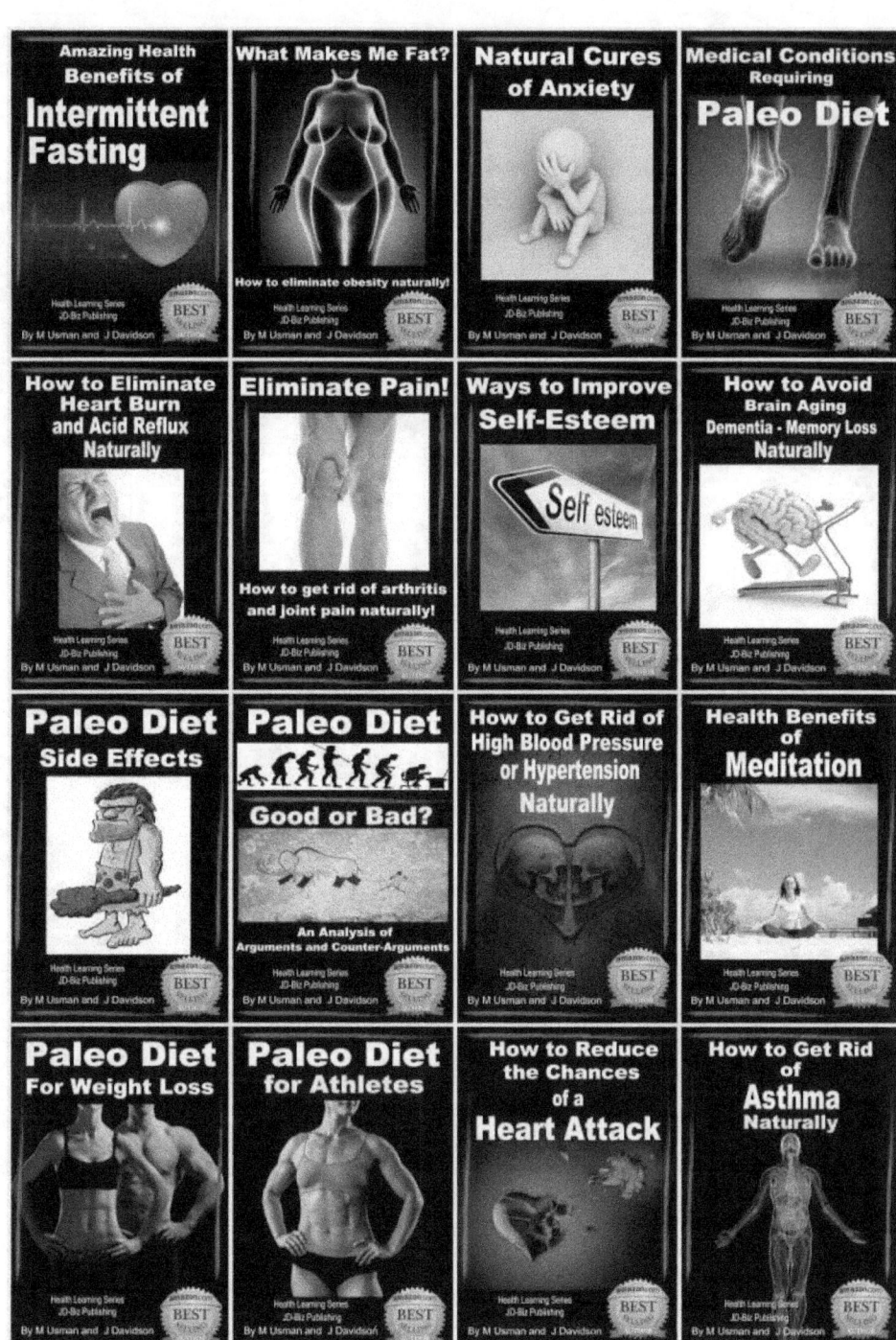

Amazing Animal Book Series

Learn To Draw Series

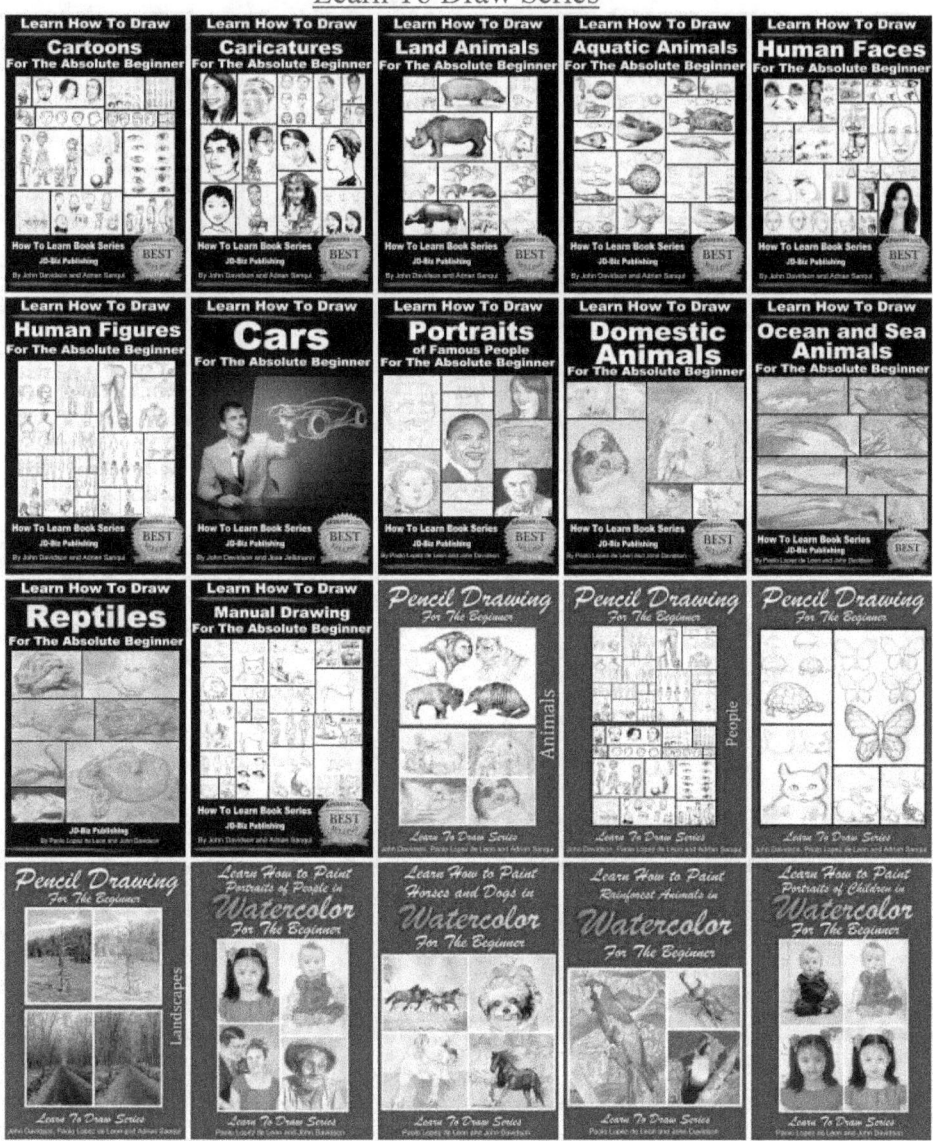

How to Build and Plan Books

Entrepreneur Book Series

Our books are available at

1. Amazon.com

2. Barnes and Noble

3. Itunes

4. Kobo

5. Smashwords

6. Google Play Books

This book is published by

JD-Biz Corp

P O Box 374

Mendon, Utah 84325

http://www.jd-biz.com/

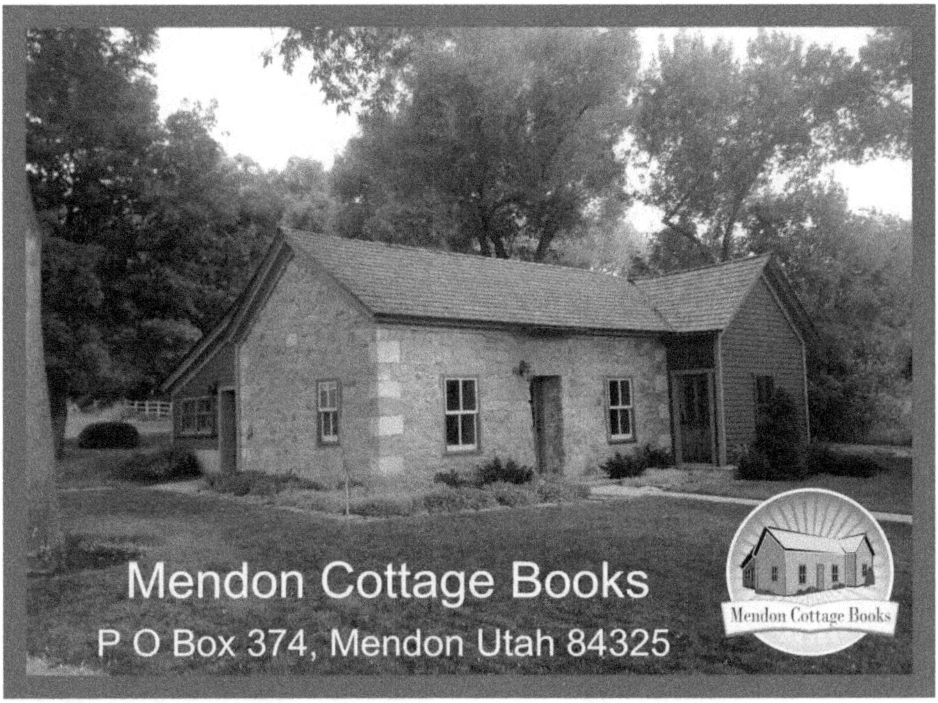

Mendon Cottage Books

P O Box 374, Mendon Utah 84325